UNFORGETTABLE
BEHAVIOR

UNFORGETTABLE
BEHAVIOR

Rosamund Kidman Cox

FIREFLY BOOKS

THOSE UNFORGETTABLE MOMENTS

Photographs catch our attention in all manner of ways, but to hold our gaze, a picture has to fascinate – to offer a story and even to make us puzzle over it. That's what the pictures in this collection do. All of them have received accolades in past Wildlife Photographer of the Year competitions. Some are famous, others not quite, but all have stories, whether of the moment itself or the fascinating lifestyles that are the background to the behaviour being shown.

It may seem that many of the unforgettable moments are a matter of luck. But few photographs of wildlife are lucky shots, especially when it comes to animal behaviour. To be able to catch the instant that makes a memorable image requires experience and fieldcraft as well as camera skill, knowing where to go and when, anticipating the behaviour and seeing the composition. For true wildlife photographers, the process becomes instinctive, so that when serendipity does happen, they have a chance of catching it.

With many of the shots you will see here, great patience has been involved. It may even have taken several years and many attempts to capture the behaviour – and to show it with artistry. The other ingredient is a love of and fascination with nature, and so often that is what shines through the photographers' work.

Leafcutter ants at work, Costa Rica.
By Bence Máté.

Previous page: Polar bears waiting for the ice to form, Hudson Bay, Canada.
By Thomas D Mangelsen.

First page: A burrowing owl grooming her chick, Pantanal, Brazil.
By Bence Máté.

ANT RIDER
Bence Máté

This is a beautifully simple picture of a Herculean
act – just one part of a complex story, enacted every
day in a rainforest in Costa Rica. The leaf-cutter
worker, classed by its size as a media (as opposed
to a minim, a maxima or a soldier), is ferrying a leaf
section back to the colony nest along an ant highway.
She (all workers are female) has cut the leaf section
from a young tree or possibly a treetop – young leaves
are favoured, being juicy and lighter – and is part of
a convoy of ants bringing a leaf harvest to feed the
fungus farm underground in the nest. The fungus is
the colony's main food, and it is fed with a mulch
made from vegetation. The more productive the farm,
the bigger the colony can grow. The photographer has
lit the scene from several angles, choosing to show the
tiny rider – a minim worker – as a silhouette. She is
a mere 2 millimetres long and has hopped aboard just
as the leaf was cut, the media worker having signalled
that she was ready to leave by stridulating – rubbing
parts of her body together to create a vibration.
The minims have many jobs, but one is to fend off
the parasitic flies that patrol the ant highways and
try to lay their eggs on the media workers. But this
scene was shot at night, when the flies aren't active,
which points to another minim job: cleaning the leaf
section of micro-organisms that might contaminate
the fungus farm. It's all part of the complex division
of labour that exists in the million-strong colony of
the leaf-cutter ant *Atta colombica*.

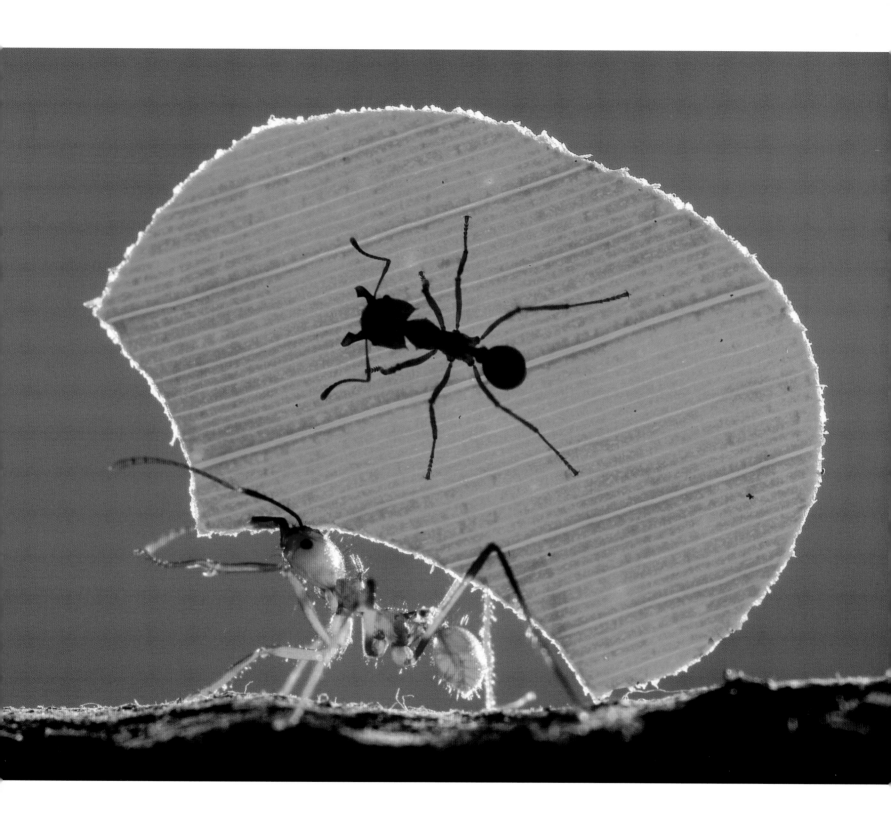

PETAL PROCESSION
Adrian Hepworth

The photographer has used a two-second exposure with a flash at the end to portray the sense of movement of the petal procession in the rainforest of Costa Rica's La Selva Biological Station. The petals have been taken from an almendro tree, and transporting each one is a worker leaf-cutter ant. On top of some segments are much smaller ants – members of the colony's minim caste. These maintain the scent trail back to the nest as well as cleaning the petals or leaves being carried and guarding the larger workers against the attention of parasitic flies. The procession is using one of the almendro tree roots as a highway over the leaf-litter – part of the route laid by the ants that morning. The petals are a seasonal bonus food for the ants. Lightweight and succulent, they make a perfect mulch to feed to the colony's underground fungus farm. In fact, depending on what's flowering in the rainforest, more than 40 per cent of material collected by the ants can be petals.

WHEN HARES GET TOGETHER
Jim Brandenburg

It's autumn on Ellesmere Island in the Canadian High Arctic, and a large group of Arctic hares are resting together on a hill. This is the first ever photograph taken of such a gathering, by Jim Brandenburg in 1986. His interest was mainly the island's Arctic wolves, but their survival is linked to that of the Arctic hares, which form a large part of their diet. The huge gatherings of hares, sometimes hundreds strong, is almost certainly a defence against the wolves – providing safety in numbers, like the flocking of birds. Only in the far north do Arctic hares behave like this, and only there do they keep their white coats all year (as do the wolves) – the summer is so short that it's not worth the energy to change colour. But once the snow melts, they become white targets against the tundra. Being one of many hares means there is less chance of being singled out. As Jim observed, 'When the wolves approached a grazing herd, the hares behaved at first like a flock of birds, moving in synchrony. Then they would scatter in every direction.' In summer, the gatherings are composed mainly of young hares, whose coats are in the process of turning white. They are easier prey than the bigger, faster adults, but become increasingly agile as they mature. Here the hares are probably a mix of adults and first-year youngsters. Once the snow arrives, they will be camouflaged, but then so too will the white wolves, all the better to sneak up on them.

THE ICE-HOPPERS
Maria Stenzel

Using a combination of crampon-like claws, sharp beaks and balancing flippers, chinstrap penguins scale an iceberg. It's a safe spot to rest and preen their feathers, out of reach of any leopard seals in the vicinity. That their feet don't freeze on the ice is because, like all penguins, they can control the blood flow to their extremities and have a counter-current system of heat exchange at the tops of their legs that keeps their feet just above freezing. The iceberg is in the Southern Ocean, close to the vast chinstrap breeding colony on the volcanic island of Zavodovski, in the remote South Sandwich archipelago, where some 1.5 million chinstraps gather to nest, safe from land-based predators. It's also surrounded by an abundance of the krill and small fish that the chinstraps need to feed to their young, which have to fledge before winter sets in. The penguins were able to rest on the iceberg because it was an unusually calm morning in the southern summer, which also allowed the photographer, on a small yacht, to take her shot. Normally, the sea off Zavodovski is notoriously rough, being in the path of westerlies and regular storms that sweep in without being tempered by any landmass.

THE MIRACULOUS MONARCHS
Axel Gomille

A water seep in the mountain oyamel fir forest of El Rosario, central Mexico, has attracted thousands of monarch butterflies. Their search for water has been triggered by a slight temperature rise in mid-March, waking them from their winter rest, which they spent roosting on the conifers, clustered in their millions. This is one of their main overwintering sites in Mexico – all mountain forests on southwest-facing slopes that offer a stable temperature and protect possibly 100 million monarch butterflies from cold, rain and predators. The monarchs arrived here in October and November, having flown up to 3,000 kilometres (1,865 miles) from the US and as far north as Ontario, Canada. Over four months, they have lived on fat stored in their abdomens. Now, on sunny days, they have begun to move down the mountain in search of water. Some have started to mate. When wind conditions are right, waves of them will begin their migration north. Those arriving in the Great Lakes region will be the offspring of migrating monarchs that laid eggs in the Gulf Coast and central regions of the US, but a few will fly all the way back from Mexico in one go – a phenomenal physical and navigational feat. In the 1990s, the population of migrating monarchs was estimated at a billion. But today, the combined effects of logging in Mexico, loss of nectar plants needed on migration, insecticide use and decline of their caterpillar food plant milkweed (mainly through herbicide use) have caused their numbers to plummet by 90 per cent in the past two decades.

THE LONG DRINK
Bill Harbin

A male giant swallowtail butterfly sucks up moisture from damp gravel beside a stream in Mexico. Its long proboscis – a food tube, formed from the left and right halves of a set of mouthparts zipped together – has unwound to reach the ground. The outer cuticle layer of the proboscis contains a rubbery protein, and when the muscles that have caused the proboscis to extend relax, it will spring back into its coiled position at the base of the head. This swallowtail was totally absorbed in the process of siphoning up the liquid, ignoring the camera just a short distance away, and continued to pump up moisture for more than five minutes. Found throughout most of North America and into parts of South America, the giant swallowtail is relatively long-lived – months in the case of males. And it is the males that are most often found puddling (the term for non-nectar feeding by butterflies) on wet soil, sometimes where an animal has urinated, on rotting material and even on carrion or dung. They are not just after moisture – in fact, after a certain amount of liquid has been pumped up, it may be squirted out of the butterfly's rear. They are after minerals, particularly sodium, and possibly amino acids, which they need in addition to the nectar they get from flowers. This male giant swallowtail may well be accumulating nutrients to pass on with his sperm to the females he mates with – a gift to help the survival of the fertilized eggs that his mate will lay.

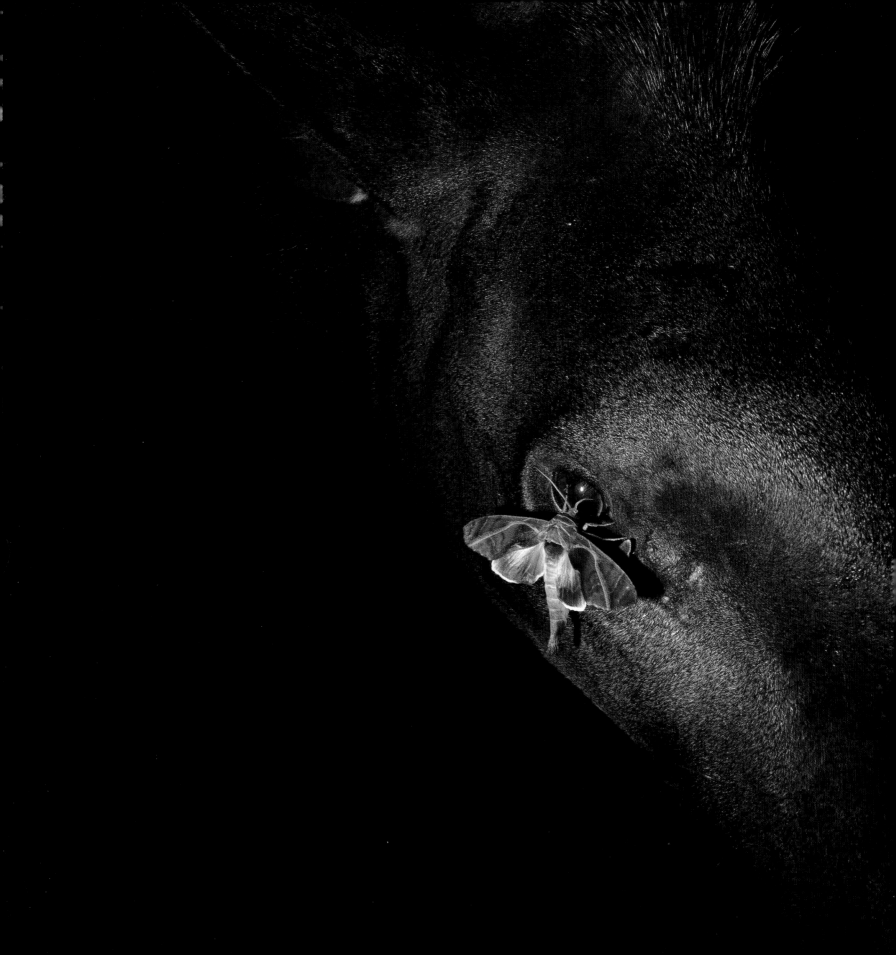

GETTING AN EYEFULL
David Herasimtschuk

At night, in the tropical forests of South America, a new cast of animals emerges to feed. Here, in an Ecuadorian swamp-forest, a moth is seen drinking from the tear ducts of a foraging lowland tapir. Whenever the tapir stopped to feed or rest, moths would try to drink from its eyes. Many different tropical and subtropical moths are lachryphagous (tear-drinking), all of them after minerals, chiefly sodium; but with mouthparts (the proboscis tube) designed for nectar-drinking, they can only obtain it in liquid form. Sodium is essential for nerve and muscle development, but as most tear-drinking moths are males (as here, showing its sexual-scent-producing glands extending from its rear), the likelihood is that the minerals are passed with their sperm to females to nourish the eggs. These moths often specialize in drinking from particular species, usually hoofed mammals, perhaps because of the chemical composition of the tears or just because hoofed animals are more tolerant. Some moths will even use their proboscis to irritate the eye so it secretes more tears; others have perfected a technique of inserting it under the eyelid of a sleeping mammal or even, in one or two cases, a bird. A few longer-lived moths – females as well as males – are known to be able to digest the small amounts of protein in tears. Many lachryphagous moths will also sip other bodily secretions, and a few will even drink blood.

THE THOUGHTFUL BABOON
Adrian Bailey

In dry-season tropical areas, waterholes, seeps and streams are where the action is, especially in the early morning and at the end of the day. Most birds and mammals need to drink regularly – in the case of doves and chacma baboons, every day. Here an adolescent male baboon, arriving ahead of his troop at a seep in Zimbabwe's Mana Pools National Park, has found the body of a Cape turtle dove. Flocks of Cape turtle doves descend every morning to drink here, and they attract predators, such as lanner falcons and goshawks, which line up on nearby trees. This dove had probably been hit by a falcon, which abandoned its prey when it was disturbed by the approaching troop. As well as plant material, baboons eat carrion and small mammals, but rather than grabbing the bird as a food item, as any other carnivore would have done, the youngster held it delicately for a good minute, turning it around, smelling it and, according to the photographer, 'gazing at the body as though in deep thought'. Baboons are, of course, highly social, forming strong bonds with each other, and very inquisitive, interested in novel objects – in other words, intelligent. There is no reason to suppose the thoughts were not unlike those of a curious human primate coming across a still-warm body. A bushmeat-hunting human would almost certainly have ended up eating the bird, just as the baboon finally did.

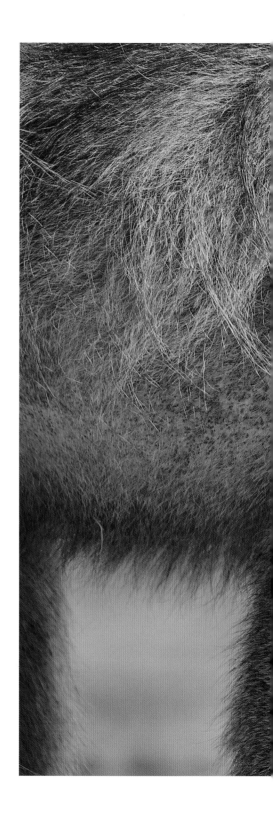

THE PENGUIN PROTECTOR
Linc Gasking

A plucky Adélie penguin attempts to drive off a giant petrel menacing a crèche of youngsters. The chicks have gathered together for protection while their parents are at sea finding food. If a giant petrel or skua threatens them, any adults nearby will try to drive it away. In the photograph, the giant petrel – a rare white morph of the normally brown southern giant petrel – appears smaller than it actually is because of the perspective. It is really larger than the squawking, flipper-waving Adélie and is capable of dragging off and pecking to death a young penguin. Usually, though, southern giant petrels act like vultures at the penguin- and seal-breeding colonies, feeding on the carcasses of the many young that perish. But when carrion is not available, they will pick off any young chicks that become separated from their parents or an older chick at the edge of a huddle. These defensive crèches form when Adélie chicks get to be 20–30 days old and both parents are at sea, and the size of a crèche depends on how many adult neighbours are around to drive off predators. At this age, the chicks are relatively safe from their mainland predator, the south polar skua, which is smaller than a giant petrel and will seldom attack a large chick. But the chicks' problems aren't over even after they fledge. When they enter the sea for the first time, they have to run the gauntlet of leopard seals that lurk offshore, waiting to pick off the unwary.

MOTHER'S LITTLE HEADFUL
Udayan Rao Pawar

It's early morning in central India's Chambal River, and a large female gharial is hanging in the water offshore. The babies that have swum out to her are using her huge head as an island, taking advantage of the safe basking perch. The female is one of a colony of gharials that have nested communally at this traditional riverbank site. The mothers stood guard over the nests and then, alerted by the grunting of the emerging babies, excavated them when the eggs started to hatch. But now they are keeping watch from the river, with this female acting as chief guard for 100 or so hatchlings. She and the other mothers, and a few males, presumably fathers, will protect the young for at least a month or more until the monsoon rains arrive and they move down the river to feed in deeper water (floodwater may also sweep the young downstream). They will keep watch for predators such as jackals and monitor lizards after the young, though unlike other crocodiles, they can't raise their bodies off the ground and give chase. Instead they have to push and slide themselves ashore. The Chambal River is the last stronghold of this fish-eating crocodile, once found in rivers throughout the Indian subcontinent but devastated by past hunting and then irreversible changes to their rivers caused by dams, canals, irrigation and other developments. Now critically endangered, gharials are still under pressure, mainly from illegal sand-mining in their nesting areas and illegal fishing, but also from egg collection, pollution and water extraction – they need unobstructed, free-flowing rivers. Gharials don't mature until they are 10 years old, and today, of the estimated 1,400 adults that survive, the most important breeding population is in the Chambal River. So these youngsters are indeed a precious little headful.

THE STAMPEDE

Eric Pierre

Muskoxen stampeding over flat, frozen tundra, thunder towards the photographer. This is not normal behaviour for these Arctic animals, unless a pack of wolves is chasing them. With their extreme insulation – dense underwool and a thick blanket of long hair – they can overheat if forced to run far, and their usual defence is to face a predator. If a lone bull is threatened, it may charge, and bulls have been known to kill people who encroach on their space. But as social animals, the more common reaction of muskoxen to a predator is to group together and face it in a line. Both sexes have upwardly curved, pointed horns, used for impaling and throwing attackers, whether wolves or bears. If they are attacked by a pack of wolves, the group forms a compact circle, all facing out and with calves in the centre. Larger adults may charge out to try to head-butt or hook a wolf and trample it (males have particularly broad 'bosses' above their horns that absorb the shock of a head-clash). But what the wolves want to do is to spook the muskoxen and make them run, knowing that the younger and weaker animals will trail behind and can be picked off by the pack. In this case, the photographer was following Arctic wolves on Canada's Victoria Island. Being careful not to spook the muskoxen, he had deliberately made a detour upwind and was half a kilometre (a third of a mile) away when they started to run, unaware of his presence ahead. Unfortunately, though, the most usual cause of a stampede is a human one – a low-flying plane or helicopter or a skidoo.

A TALE OF TWO FOXES
Don Gutoski

This is a rare picture of a rare event – an Arctic fox killed by an American red fox. But it is one that may occur more often as the climate gets warmer and the larger reds move northward into Arctic fox territory. The photographer was in Canada's Wapusk National Park, on Hudson Bay, when he saw from a distance a red fox chasing something. By the time he was close enough to see what it was, the smaller fox was dead. The photograph shows the red fox, having eaten much of the body, starting to drag the remains away to cache for later. It's unlikely it had set out to hunt the Arctic fox – indeed the two species normally avoid each other. But both feed on lemmings and other rodents, and so where their ranges overlap, they can come into conflict, especially when snow lies on the ground (it was -30°C (-22°F) when the picture was taken) and finding food is a struggle. In future years, as the Arctic continues to warm and tundra is replaced by forest, the Arctic fox will lose more habitat and red foxes will expand further north. Not only will reds kill the smaller foxes, they are superior hunters and so will outcompete them. Milder and shorter winters will also reduce the supply of lemmings that inland Arctic foxes depend on. So their outlook is bleak, unlike that of the ever-adaptable American red fox.

BALANCING ACT

Joel Sartore

On a sheer cliff face, a female mountain goat performs what seems like a death-defying feat but is just part of everyday behaviour for these mountain-climbing champions. The photographer anticipated the procedure, knowing that mountain goats often climb down the cliff in this gorge in Montana's Glacier National Park to access minerals seeping out of the rock. Positioning himself on the other side of the gorge, he watched the balancing technique step by methodical step. This North American species is not actually a goat. It's a member of a group of animals that include mountaineers such as the European chamois and the Asian serow. For climbing, it has narrow, cloven hooves with pointed toenails surrounding flexible, textured, protruding traction pads. It also has muscular shoulders and a thick neck that, once it has hooked onto a ledge, help it pull itself up vertically. In this instance, the animal first balanced on all four hooves on the tiny ledge opposite the mineral-licking area and then pushed out with her front legs to wedge herself into the crevice. When she'd finished licking the rock, she reversed the move with perfect balance, pushing back onto the tiny ledge. Then she slowly pivoted around and climbed back up the cliff, pressed against the wall, moving almost vertically, and seemed to lift herself up with sheer strength, edging to one side when she couldn't find a foot-hold straight above. The sheer inaccessibility of the mountain goats' barren, perpendicular world keeps them safe from most predators, though eagles sometimes take kids. But their mountain refuges are also the greatest cause of deaths, from avalanches in the spring, rockfalls and winter starvation.

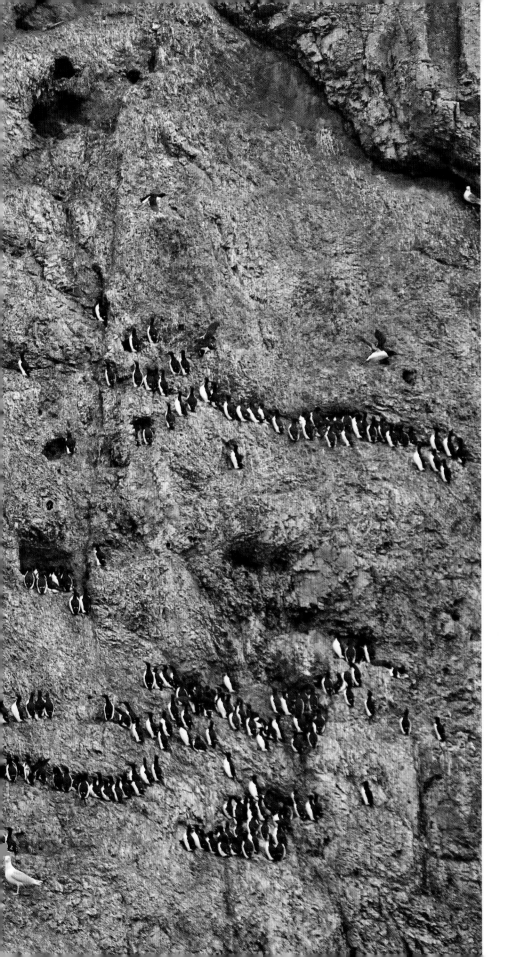

PERILOUS PICKINGS
Jenny E Ross

It's a desperate polar bear that decides to climb down a cliff for a scanty meal of scrambled eggs. Here a hungry young male clings onto a ledge, scavenging the eggs of a colony of Brünnich's guillemots. Glaucous gulls, the only predators that usually assail the guillemot colony's high-rise fortress, hang around waiting to eat any eggs he misses. The polar bear may spend more many hours scavenging on the cliff face. It's hard work for a huge animal not designed for rock-climbing. Not only will he expend a lot of energy for meagre rewards, but he will also risk falling into the sea below. A lack of sea ice drove him onto this island in the Novaya Zemlya archipelago, part of the Russian Arctic National Park. Historically, the sea ice to the far north and east of these islands has remained frozen in summer. But today the ice is melting earlier and receding further, forcing polar bears on shore. The result is an increasingly long summer and autumn fast before the sea refreezes. Ice provides the polar bears with access to the fat-rich marine mammals, the seals, that they depend on to build up the energy reserves needed to survive the lean months. If current climate trends continue, scientists predict that, within just a few decades, ice-free periods in some Arctic regions will be too lengthy for polar bears to survive.

BEACHED BEARS
Howie Garber

A congregation of polar bears. They are not looking out to sea but feeding, though with an eye out for the waves. The feast is provided by a grey whale, killed a week earlier by killer whales and washed up on this spit of land near Point Barrow off the northern Alaskan coast. Glaucous gulls have also flocked to the beach to feed on the bounty. More than 30 bears are gathered at the site – in vision are an adult female, an adult male, subadults and cubs, with the rest not far away – and another 50 bears are within a couple of miles. Bears often congregate along this coast, attracted by the remains of bowhead whales that the Inuits hunt (a butchered carcass can be seen in the background, part of an annual quota), but normally nowhere near in such numbers. The sea ice has melted, and what was left has been blown by a storm miles off into the Chukchi Sea. So the bears are stranded until the sea freezes again. This gathering, in August 2002, was exceptional, but the retreat of the sea ice from the coast is not. Though this very northerly Chukchi population, spanning Russia and Alaska, is not experiencing such a great percentage loss of ice as most other polar bear populations, access to their sea-ice habitat is gradually being reduced. Also, though land-based food such as carrion, eggs and berries can prevent starvation in times of no ice, studies show that, for their long-term survival, these huge bears still need their specialized diet of fat-rich ice seals for at least half the year.

BODY BUILDING

Christian Ziegler

It's late afternoon in the rainforest on Barro Colorado Island, Panama, and the first column of *Eciton hamatum* army ants has started to create a new nest out of their bodies. Using their strong claws and mandibles to link together, they are forming chains and nets, layer over interlocking layer, which will build into a living bivouac, with chambers and galleries. Eventually, all 300,000 ants in this colony will arrive, the last workers carrying the eggs, larvae and queens from the previous night's nest into the centre of the new one. They transport them after dark to avoid the attention of parasitic phorid flies, which would try to lay eggs on them. During the night, the bivouac can change shape, adjusting the surface-to-volume ratio and therefore the amount of heat loss, to keep the internal nursery temperature stable. These ants are nomadic hunters specializing in raiding the nests of other social insects, in particular, ants, bees and wasps, and at very first light, a continuous column of raiders will leave the nest. It will fork into smaller columns, searching for insect food to bring back to the colony. To enable the raiding swarm to move efficiently across the forest floor or between twigs and branches, they hold claws to form living bridges. There is no lead ant – calculations are made as a unit through 'swarm intelligence'. And at the end of the day, once a new bivouac site has been located, the ants start to pour into the new structure, and the old one will dissolve into a mass of workers, once again carrying the queens and young to the new location, where the whole mass will rest until first light.

SIZING UP

Klaus Tamm

Comparatively little is known about these long-legged neriid flies – or indeed, about the behaviour of very many insects in the tropics. And so this photograph is both special aesthetically and as a record of a seldom-witnessed encounter. These stilt-legged males, just 1.5 centimetres ($1/2$ of an inch) high, were part of an aggregation of neriid flies that had been attracted to fresh gecko droppings on the veranda of a house on the Indian Ocean island of Réunion. They were feeding on the droppings, and every so often a couple of male flies would break off and, as the photographer describes, 'engage in a kind of combat dance that involved spinning around each other, while striking out with their forelegs'. They would finish by stretching up to their full height, then would push with their mouthparts, shoulders and forelegs until one was higher than the other. Male neriid flies have much longer legs than females, presumably for use in these combat rituals. Other neriid species that have been studied fight over access to females that are ready to lay eggs. And when a male has mated with a female and she starts to lay her eggs, he will enclose her within the span of his legs to prevent rival males from getting to her.

BIG CAT FIGHT

Andy Rouse

It was the noise that was terrifying, says the photographer, 'the most violent, amplified cat-fight yowling you've ever heard'. This was no play-fight between adolescents but a major fight between mother and daughter. The mother (left), heavier and stronger, clawed her daughter and knocked her to the ground, whereupon the daughter fled. The mother was Machaili, one of the most famous tigresses in India, certainly the most photographed and a star of Ranthambore National Park. Her fame came from her hunting prowess, her strength (among other feats, she was filmed killing a huge crocodile) and her success at protecting her cubs. Indeed, some male tigers were said to be frightened of her. Though a mother will tolerate grown offspring in her territory for a while, if she has a new litter or food becomes scarce, the likelihood is that she will drive off competitors, which in this case included her daughter Satra. Eventually, though, Satra would oust her mother and take over all Machaili's territory. In 2016, aged 20, Machaili is the oldest-known wild tigress. She has lost her canines (in a fight with a crocodile) and is forced to hunt smaller prey or drive leopards off theirs. But her prowess lives on in her progeny, which now populate not only Ranthambore but also Rajasthan's Sariska Tiger Reserve, where two females have been moved to found a new dynasty there.

THE EGG-SNATCHER

Mark W Moffett

Barely 2 millimetres long, its legs held close to its body, the tiny brown spider could be a bit of windblown detritus or even a dead insect. Certainly *Epus*, the much larger Sri Lankan jumping spider, is unaware she has an intruder in her nursery web. That intruder is *Phyaces*, also a member of the jumping-spider family but one that specializes in hunting spiders – other jumping spiders. This seems a particularly dangerous lifestyle as, like *Epus*, all jumpers have all-round vision, with two of their six eyes forward-facing and exceptionally large, all the better for hunting focus. That *Phyaces* has managed to enter the nursery nest unobserved is because she (and this one is a she) has moved in very slowly, with short, barely noticeable steps, using a rocking gait, perhaps to mimic a bit of dirt rolling in the wind. Once close to the eggs, she stays motionless, palps and legs pulled in, until the time is right to grab an egg. Holding it, she appears to perforate it and then suck out the contents. (Normally, a spider has to liquefy its food, using its fangs to inject venom to pacify its prey, then forcing digestive fluid from the stomach into it or vomiting enzymes over it before chewing or sucking up the contents.) The likelihood is that *Phyaces* will risk staying in the nursery to feast on the remaining eggs, sucking them like energy drinks, and will never be caught by *Epus*.

SALMON SWIPE
Paul Souders

In a pool below rapids, along a small river running down to Kuliak Bay in Alaska's Katmai National Park, a female grizzly bear fishes for salmon. The salmon are gathering before attempting to leap the rapids and continue upstream to spawn. In late summer, most of the grizzlies in this southern Alaskan wilderness area congregate along the rivers to feast on the salmon returning from the Pacific to their natal streams. This female was experienced at fishing, using different techniques depending on the depth of the water. Here she is swimming above a shoal of chum salmon, trying to scoop them out with her massive clawed paws. Many grizzlies in Canada and Alaska depend on the autumn run of fat-rich salmon to put on enough weight for winter hibernation and, for the females, to build up enough reserves to bear young and suckle them through the spring. Though grizzlies can survive on a mainly herbivorous diet, those bears that have access to salmon need smaller home ranges, gain weight faster, grow larger and rear more young. The ecosystems alongside the rivers are also moulded by the salmon, which fertilize the soil through the scraps and dung left by the bears that feast on them.

BEAVERING
Louis-Marie Préau

In the River Loire in France, a European beaver swims to the underwater entrance of its lodge carrying a poplar branch and a mouthful of vegetation for its family inside. The river here is relatively deep, and rather than building a lodge, the beaver couple have dug one into the riverbank under the roots of a poplar. The tree roots act as props for the gallery tunnels and for the main family room, which is above the water level. Had the water been shallow – in a stream they'd dammed, a lake or a pond – the beavers would have constructed a lodge using logs, branches, stones and mud, and with its entrance under water. Normally, beavers forage at night, but if undisturbed, they will be active in daylight, as here. Capable of holding its breath for 10–15 minutes, a beaver can close its nostrils and ears, has a transparent eyelid (a nictitating membrane) and has lips that close behind its chisel-like incisors so it can use its teeth under water. In winter, the family won't hibernate, but where the winters are cold, they will have an underwater food store of twigs and branches so they can stay in the lodge when the water freezes over. This beaver is descended from a group that was reintroduced to the Loire in the 1970s. The original population was hunted to extinction, as were most European beavers, mainly for their fur, but also for the tail glands that keep their double-layer coats waterproof. Now at least 24 European countries have reintroduced beavers. Among the few western ones still to do so officially are Italy and mainland Britain.

BIG CHIPPER

Benjam Pöntinen

With a sideways flourish, the black woodpecker tosses out woodchips from the nest hole he has been fashioning. He has been at it for the past two weeks, since early March. The ground below is littered with large woodchips – an obvious sign that Europe's biggest woodchipper is at work here in Finland. The hole – 7 metres (23 feet) up the pine tree – is a major excavation, probably extending 60 centimetres (2 feet) down into the trunk. The amount of head-pounding hammer-drilling involved would cause significant trauma to the brain if the woodpecker's head didn't include elasticated tissue between beak and skull, spongy shock-absorber skull-plate bones and a special hyoid bone that diverts vibrations and acts as a safety belt. Its chisel-like beak has a high-strength inner layer of bone and a flexible outer layer that helps reduce the shock of the vibrations and keeps growing so the beak doesn't wear down. Bristles keep the sawdust out of its narrow nostrils, and thick nictitating membranes close over the eyes at the moment of hammering. If the female finds the nest chamber to her satisfaction, she will lay two to eight eggs, which the pair take turns incubating (at the changeover, they will communicate by drumming, inside and outside the hole). The oval entrance hole is a perfect fit for the woodpeckers, but it is also large enough for a pine marten to squeeze into and eat the eggs, chicks or even a brooding woodpecker. The risk of a pine marten returning to a nest hole the following year may explain why a black woodpecker usually starts a new one every year.

LIGHT PATH

Charlie Hamilton James

In a flash of light, a kingfisher flies straight to his nest hole with a minnow for his chicks. His mate – just visible perched on a branch downstream – is digging a new nest for a second brood. The territory that the pair has held for several years is on a river in southwest England, which the photographer knew intimately. But to take this photograph was a challenge. Kingfishers are small and shy and move very fast, and the nest hole was nearly 2.5 metres (8 feet) above the river on a vertical riverbank. It was above flood level and safe from predators but a difficult location to light in a way that would reveal the bird, bank and background. The kingfishers take about 25 days to rear the brood if there is a good supply of fish, working from sunrise to sunset (in this case from mid-April to mid-May), catching more than 1,000 small fish (minnows are a favourite – easy to manipulate and swallow). The pair's choice of nest site and territory is therefore vitally important, with good fishing perches, where the water is relatively shallow and clear enough to see the little fish they prefer to catch. Once the young have fledged, they will be tolerated for a couple of weeks and then driven out of the territory – in this case, so the pair could concentrate on rearing a second brood.

A KINGFISHER IN HIS CHAMBER
Angelo Gandolfi

A male kingfisher takes his turn brooding the clutch of seven almost perfectly round, glossy white eggs on a nest lined with regurgitated pellets (fish bones and scales). He has positioned himself exactly on top of the eggs and is rotating them one by one to make sure they are incubated evenly. The tunnel is an artificial one, constructed by the photographer on the River Scrivia in Liguria, Italy, after the kingfishers' riverbank nest of the previous year had been destroyed. Glass fitted in the wall of the nest chamber and a person-sized burrow created in the bank behind it allowed photographs to be taken. Though the tunnel had to be kept in darkness, the kingfishers paid no attention to the occasional flash or click of the shutter, seeming to feel completely safe in their dark cave. The pair – identical except for the orange-tinted bottom half of the female's beak – took turns brooding the clutch. When one bird arrived at the nest entrance, it would call the other out. On this occasion, before entering the tunnel, the male gave the female a present of a small fish. Once the eggs hatched, 20 days later, the pair took turns bringing fish for the chicks, though the male would do more of the fishing and the female would brood them overnight. That first year, the pair reared two clutches of seven eggs in the artificial tunnel and continued to do so for several more years.

BEAST OF THE SEDIMENT

Göran Ehlmé

A big male Atlantic walrus ploughs the seabed off Greenland, sucking clams from their shells. Surprisingly, such a big, blubber-insulated animal – a male can weigh more than 1,450 kilograms (3,200 pounds) – can survive on a diet of mainly bivalves (clams, cockles and mussels) and other small bottom-feeders. Shellfish in particular are rich in nutritious fats, and a walrus is a fast, efficient forager and can hold its breath under water for up to half an hour. Here it is using its right flipper (walruses tend to be right-flippered) to remove the top layer of sediment and uncover bivalves as it ploughs through the mud. Moving its tongue like a piston, it sucks out the soft tissue of the shells surprisingly fast: a walrus has been recorded eating more than six clams a minute, and one may well eat more than 50 in a dive. Rather than flipper-ploughing, it may also squirt jets of water out of its mouth to clear the sediment ahead; or it might root through the sediment with its muzzle. The technique depends on the density of prey, the light and the clarity of the water. Normally a walrus feeds in the day, and probably sees as much as it feels, but it can feed at night, too. Though its eyes are small, they have a tapetum lucidum – a reflective layer behind the retina – which makes them more light-sensitive, and they can face forward for binocular vision. Here its eyes are protruding as it looks around before surfacing. Its tusks (canine teeth) aren't used for feeding but for hauling out on ice or rocks, for defence and, in the case of a big male like this one – its neck covered in thick, protective skin – for fighting.

THE PAPER-CLIP SUITOR

Tim Laman

A male great bowerbird peeks into his bower waving one of his prize decorations, a pink paper clip. This is a female's-eye view of him, framed by his wickerwork avenue – the bower – which he will entice her to walk down. A few green seeds on the walkway hint at the green and white artistic display laid out on the court area at the other end of the bower. He has a preference for green and white, but strategically placed among the arrangement of glass, snail shells, pebbles and seeds are his special pink and red items – bits of plastic that he has collected on the campus of Townsville University, in Queensland, Australia. If a female is impressed enough by the male's enticements to enter the bower avenue, he will mesmerize her with a peekaboo of various prize objects and a flash of pink by raising his head crest. Accompanying this will be a repertoire of calls, many mimicking those of other birds. Even lighting is manipulated – the bower is orientated across the path of the sun. If a female is suitably impressed by his bower, his artistic skills and his posturing – more so than his neighbours' colour schemes and creations – she will allow him to enter the bower and mate with her. It will be a brief affair, and she will then leave to rear a family on her own. She doesn't need a partner: the hatching of the chicks will coincide with the annual northern Australian monsoon and a resulting bounty of food. The male, meanwhile, will continue to devote his time to bower maintenance and improving his display of treasures in the hope of attracting more females. Females favour the more experienced builders (bowerbirds can live for more than 20 years), often the ones who learn how to steal the gems from the bowers of nearby younger males to enhance their own.

HEADSTRONG HELLBENDERS
David Herasimtschuk

Two male hellbenders, jaws clamped together, battle it out, twisting and turning as the fast current pushes them downriver. They are head-to-head over prime breeding territory, centred on an ideal nest space – a cavity under a large, flat stone in the middle of a river. It's a cavity into which a territory-holder will entice a female – sometimes more than one – to lay her hundreds of eggs, which he will then fertilize before driving her out. He will guard the nest site, keeping the eggs relatively free of pathogens and fending off potential predators. The winner of the fight between this pair was the larger, territory-holding male. Though they look like swamp monsters, with their wedge-shaped heads and wrinkled, slimy skin, these giant North American salamanders – up to 74 centimetres (nearly 2½ feet) long – are found only in fast-flowing, oxygenated water. When submerged, they breathe through their skin, and the wrinkles increase the skin's surface area. Declining water quality and silt and sediment run-off from the land, along with collecting of these slow-maturing, long-lived amphibians for the illegal pet trade, has caused a 80–90 per cent population crash. Today they are mainly confined to Appalachian streams and rivers – a symbol of crystal-clear, healthy water.

A STACK OF SUITORS
Marcel Gubern

For most of their long lives, green turtles are probably solitary, and it can be 20 to 50 years before they are mature enough to want to come together to mate. Then, every few years – the timing dictated by the day length and sea temperature – they return to the coast where they were born, in this case, Sipadan Island, off the east coast of Sabah, Borneo – a hotspot for turtles. The males usually arrive first, and the females follow over a few weeks, though they are receptive for only a few days. So when there are more males than females, as is usual, the competition is intense, and mating can end up being brutal. The hormonally driven males may even mount other males or even human divers. Here, the female is carrying two males. The first male is mating with her, holding on with his flippers and with his tail hooked under the back end of her shell, his long penis firmly inserted. The second male is trying to dislodge him, biting his neck and then his tail, while a third male is circling, watching for a chance to dislodge the others. The first male may well hang on for a day, which will result in a lot of exertion for the female, who has to get to the surface to gulp air every five minutes or so. The likelihood is that, once she has had her fill of mating, she will have received the sperm of at least three males. Then every couple of weeks, she will go ashore to lay up to five clutches of 50 to 200 eggs, but she won't mate again unless forced to, rebuffing the attentions of the males hanging around offshore. Interestingly, there is seldom an equal mix of the hatchlings' fathers, as some sperm, often that of the first male, seems to be better at fertilizing eggs than others or perhaps has more of a chance.

GREAT CATCH
Bence Máté

If you look at the beak of a reed warbler – narrow and pointed – it suggests what the bird eats: most likely insects, small ones. The great reed warbler, the biggest of the European reed warblers, is known to have a varied diet of varying size, ranging from small flies, beetles and bugs, to large spiders and damselflies. But the photographer was taken by surprise when he witnessed a great reed warbler swoop down and snatch a minnow. He had spent the night in a floating hide in the lake, waiting for dawn and little bitterns to emerge from the reeds to feed, so he was at the right eye-level to catch the action. There was nothing unusual about the location: a lake in southern Hungary, part of Kiskunság National Park, with reedbeds and a profusion of normal warbler prey. But great warblers are both opportunists and aerial acrobats, and this one may well have become practised at snatching washed-up insects on the surface of the lake. And with a nest in the reeds and a large brood to feed, a fish would be worth taking, or even a small frog or newt if one presented itself. Crucially, the wind was also down, allowing the warbler to skim the surface easily and to see below the surface.

THE BADGERS AT HOME
Kai Fagerström

In a little house in a wood in central Finland lives a badger clan. When the house was abandoned, the badgers moved in. They excavated the sett – their main home – below the ground floor, with an exit in the kitchen under the oven. The cubs are born there in winter. They emerge in spring, and until they are big enough to forage with the adults, they play in the house, climbing the stairs to the bedrooms, where other animals also live. Paths lead from the house to various foraging areas – pasture on the edge of the forest, where they forage for their main food, worms and other invertebrates, and a hazel stand in the forest where, in autumn, they fatten up on nuts. In summer and autumn, they feast on fallen plums and apples in the orchard and pluck gooseberries from the bushes. There are other badger setts in abandoned houses – sheltered, relatively warm locations for hibernation dens, even better than under boulders in the forest. But like all badgers in Finland, these house dwellers are very wary of humans. Hunters kill at least 10,000 badgers a year, supposedly to protect gamebirds (partridge, grouse and capercaillie). Badgers will eat any eggs, chicks or smaller mammals they come across, but their diet is mainly invertebrates, supplemented by nuts, fruits, bulbs and tubers (crops when available). In Finland, badgers are at the northernmost limit of their range, needing a certain amount of snow-free days so their cubs can put on enough fat to survive the winter. And where foraging areas are widely spaced, clan ranges have to be larger and more overlapping than, say, British ones.

WHEN THE LION LIES DOWN WITH THE FOAL

Adrian Bailey

A lion embraces a zebra foal. The foal is not injured and is being held gently. But all is not what it seems. The lion took the foal after it became separated from its herd, which was moving on migration through the lions' territory in Chobe National Park, Botswana. When the picture was taken, there was just one lioness nearby, but not long after, the rest of the pride arrived just as the foal got up and started to move away. Within minutes it had been caught and killed. The likelihood is that the lion was merely playing with its feeble young prey and had lost interest in it for a while; had the foal struggled, it might have been killed sooner. Very young antelopes that haven't yet developed the flight instinct are sometimes taken by lions and respond to their predators as protectors, behaviour which stops them being killed, at least for a while. There are, however, rare instances of lionesses taking the young of their prey animals and keeping them alive for days or even weeks. A famous case occurred in northern Kenya in 2004, in Samburu National Park, where a lioness caught a young oryx calf and adopted it, allowing it to curl up and rest beside her and protecting it from other predators. In fact, she caught calves five times, each time looking after the baby as if it were her cub, following it everywhere and even preventing it from making contact with other oryx. The first calf lasted two weeks before being killed by a lion when the lioness was asleep. The second, third and fourth were removed by rangers. The last, a newborn, finally died of starvation. The reason for such unusual, obsessive behaviour was possibly trauma – the lioness having become separated from her pride, possibly even her own cubs – and the young oryx probably gave her comfort.

GANG CULTURE
Andy Rouse

A gang of adolescent striated caracaras, on West Falkland island in the Falklands, has surrounded a gentoo penguin chick. Its parents are at sea, and it has made the mistake of straying too far from the crèche of penguin chicks and any nearby adults that might have driven off the caracara gang. It doesn't stand a chance. Caracaras are birds of prey, scavengers but also killers. Highly intelligent and highly social, striated caracaras are found in the extreme south of Chile and Argentina and on the Falkland Islands, feeding on anything they can find, whether carrion, insects, eggs or small or baby birds. Their curiosity probably helps them find food in unlikely places, including around human colonies, but their best living is to be made from the vast penguin colonies. Adult pairs hold territories around the colonies, but adolescent and unpaired caracaras have to find another way to survive – ganging up to hunt. They bully adult penguins off their nests to get at the eggs and chicks and attack bigger prey such as sickly seals or even sheep, which has resulted in their persecution in the Falklands. Within the gang, which can be more than 30 strong, there is a hierarchy, with the older, stronger birds getting the pick of the food.

FATHER PROTECTOR
Mark Payne-Gill

A male giant bullfrog rams and digs his way through thick clay mud to engineer a canal that will release water into the nursery pool. It's summer in South Africa and more than 35°C (95°F). The pool where he mated and the eggs were laid was formed by a heavy downpour of rain a little more than a week ago but is now a puddle. It was this rain that triggered his emergence (and that of all the bullfrogs in the area) from his underground burrow after months of semi-dormancy, encased in a waterproof cocoon of skin. Having paired with a female, he fertilized hundreds of eggs as she laid them in the little pool and then stayed both to guard the offspring and to engineer the water level. It's now a race against time as the puddle dries out and heats up. But shallow, warm water also speeds up the development of the tadpoles. With the right water level and temperature (they will die if the water gets too hot), a clutch might develop from eggs to baby frogs in just 17 days, with metamorphosis (the change from tadpole to frog) taking just five days. But in the meantime, the tadpoles are prey to other frogs and birds, and to guard them effectively and engineer the water regulation, a male needs to be big and strong. Indeed, unusually among frogs, male bullfrogs are more than twice the size of females. This male was nearly 20 centimetres (8 inches) long. He was also fierce and lunged at the photographer, jaws agape (he has teeth). Capable of seeing off even a heron, he will guard his offspring until they leave the water, though he might also snack on a few to keep up his strength.

THE BLUSH
Anup Shah

This is the moment of consummation, after several days of courtship. The photographer, in Kenya's Maasai Mara National Reserve, had watched the build-up and witnessed the female ostrich finally take the initiative. She began to preen herself and then approached the male, her wings outstretched with the tips fluttering, and began to dance and bow around him, her head held low, opening and shutting her beak. It had the desired effect: his neck turned crimson and puffed out. She then led him on a short run, dropped to the ground and let him crouch over her and gently drop onto her back. Mating only takes a minute, but unusually among birds, a male ostrich uses a phallus, something only 3 per cent of birds have, belonging to two ancient groups – one including emus and kiwis, the other geese and ducks. The rest reproduce through a 'cloacal kiss', which allows the female to choose whether or not to draw up the sperm or reject it. In the moments after copulation, the male ostrich sways rhythmically from side to side vibrating his wings and then gently steps off. He has already created a nest scrape into which the female will lay an average of 13 large eggs – the largest eggs in the world – which they will take turns incubating. One or more 'minor' females may mate with the male and lay their eggs in the nest until there can be a clutch of 30–70 and occasionally more, but as an incubating ostrich can usually cover only about 20, surplus eggs are pushed out by the main female, who somehow manages to reject only eggs that are not her own.

RACE FOR LIFE
Zig Koch

Bursting out of the forest, South America's largest land predator, the jaguar, sprints down the beach after the world's largest rodent, a capybara. The name jaguar derives from the native word *yaguará*, supposedly meaning 'beast that overcomes its prey in a bound', reflecting the fact that it is a stalk-and-ambush hunter rather than a sprinter. Indeed, this capybara outran the female jaguar and plunged into the river. A jaguar can swim, but the semi-aquatic capybara not only swims but can also dive, closing its ears and nostrils when heading down. But in the wetlands of Brazil's Pantanal, capybaras make up a significant part of a jaguar's diet. Pantanal jaguars are also larger than Amazon rainforest ones, with the weight and power to kill large animals, though being opportunists, they will hunt whatever prey presents itself. To kill a heavyweight capybara – big males can weigh more than 70 kilograms (155 pounds), and the average weight is 50 kilograms (110 pounds) – a jaguar uses a specific technique, made possible by its thick canines and well-developed head muscles. It stalks and then pounces, delivering a bone-piercing bite to the skull that penetrates the brain, or it grabs the capybara with a bone-breaking bite to the throat. In the case of a caiman (a crocodilian), which has a thicker skull and is dangerous, it goes for the nape of the neck and may even stalk the reptile from the river.

RESPECT
Igor Shpilenok

When it comes to territory, small predators can be fierce beyond their size. Here the photographer's cat, Ryska – her name means little lynx in Russian – stands outside their cabin and, with aggressive posturing and growling, warns off a fox. In winter, searching for food, foxes would regularly visit the cabin in Kronotsky Nature Reserve in the Russian Far East. If one peered in at the window, which was possible when the snow was deep, Ryska would sit on the other side, fur raised, and growl. When she was outside, she would hold her ground. The foxes were not always frightened, and so encounters could be a sort of dance, but if a fox came too close, Ryska would probably have attacked it. In the wilds of Russia, foxes and domestic cats seldom meet, but in many urban areas of Europe, they encounter regularly at night – and mostly ignore one another. In the UK, a typical fox territory can be occupied by at least 50 cats, some of them heavier than an average-sized fox. There is little evidence that foxes ever kill cats, other than possibly very sick animals or young feral kittens, and if a fight does break out, it is almost always when a territorial cat attacks the fox, which invariably flees. In fact, it is usually fox cubs, ever curious, that end up with serious injuries from encounters with cats.

THE DUEL
Sergey Gorshkov

An Arctic fox braces itself as a snow goose attacks. The gander is defending his nest against the marauder, intent on snatching eggs. Here on the Russian island of Wrangel, north of Siberia, Arctic foxes survive on lemmings. But spring brings a short-lived bonanza. Possibly 100,000 lesser snow geese arrive to nest, having flown more than 4,800 kilometres (3,000 miles) from their wintering grounds in North America. It's the last major breeding population of snow geese in Asia. They lay eggs in early June, the number of nests depending on the snowmelt and the extent of bare ground. When nests are close together, the neighbours may help drive off raiders. But when there is more snow-free ground – as there has been in recent years with increasing Arctic temperatures – and the nests are widely spaced, the foxes might take more than 20 eggs a day, which they cache in the tundra soil for harder times. The nests they have most luck with are either unattended or with just the female present; if she leaves the nest to drive off the fox, it has a chance to nip in and snatch an egg. All the colony's eggs hatch over a week in early July. It's a safety-in-numbers strategy: there are only so many goslings a fox can kill in such a short time. If, however, bad weather means egg-laying and therefore hatching are strung out, then foxes can have an effect on the number of goslings. But once most of the goslings have hatched, the entire population leaves, the army of adults acting as protection as they walk to the wetland feeding area where the goslings will be safe.

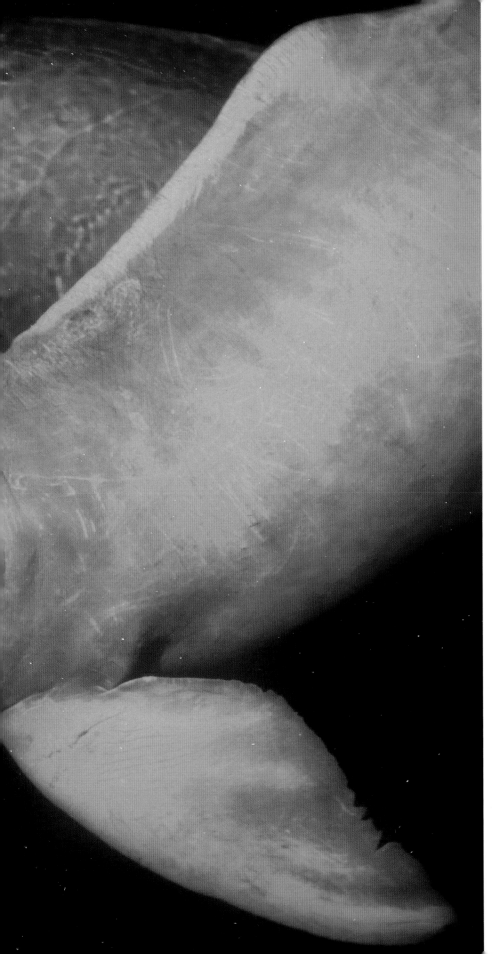

BOTO BALL PLAY
Kevin Schafer

This is possibly pure showing-off. A male Amazon river dolphin, or boto, is throwing around a macucu nut, watched by other botos, mainly males but possibly including a female. The photographer watched botos playing with objects many times, most often in the afternoon, and to shoot the behaviour from above, built a floating platform over a tributary of the Rio Negro in Amazonian Brazil. Though females have been seen playing with objects, ball-throwing seems to occur mainly among groups of adult males as a sort of aggressive competition. Any object will do, a branch, a ball of mud or even a turtle. These botos are in tannin-rich water and so appear orange, but as the nose of the ball-thrower reveals, they are actually pale-skinned, and depending on their age and sexual arousal, males can be bright pink. The largest of the world's river dolphins, botos can be 2.5 metres (more than 8 feet) long and capable of catching large fish with their long jaws, which are also useful for winkling prey out of hiding. Unlike marine dolphins, a boto's neck vertebrae are unfused, allowing it to turn its head from side to side. In the flooded forest where it hunts, this flexibility enables it to swim around tree roots – and, of course, throw 'balls'.

THE LIVING LARDER
Mike Gillam

In a chamber, a group of Australian honeypot ant workers hang like storage jars, their abdomens swollen with sugar-water. These 'repletes' may hang for months, waiting for a time when their sisters ask to be fed. Having 'major' workers (large in terms of the other workers but only about 15 millimetres, or half an inch) that can store fluids in this way is how a honeypot ant colony survives the dry periods in the semi-desert areas where they live. The nest of this Australian species, *Camponotus inflatus*, may extend deep into the earth to where the temperature is cool and stable. It will have many galleries running off a central shaft or shafts, with lots of larders. The nest is almost always associated with mulga acacia trees. These have special nodes that secrete nectar, which forager honeypot ants (from the 'minor' caste) collect, in turn providing protection from any leaf-munching insects. Mulgas may also have psyllid leafhopper insects feeding on them. Like aphids, they secrete honeydew that the ants harvest, again providing protection. When the foragers return to the nest, they regurgitate the honeydew into the mouths of the repletes. As a replete's crop fills, the flexible membrane connecting the plates that protect the abdomen stretches so much that their scales appear as dark stripes. In the light, the colour of the abdomen depends on whether it contains mainly sugar-water (sucrose) or more dissolved solids and fructose and glucose, leading to a darker shade of amber. Either way, these honeypot ants are considered delicacies by the Aboriginal people of Australia and by many wild animals, which may be a secondary reason the nest shafts extend up to 2 metres (6 feet) down into the ground.

BLIZZARD OF SEX AND DEATH

José Antonio Martínez

The blizzard is a maelstrom of more than a million mating mayflies. Standing in the swarm was, says the photographer, like 'being in a gale, brushed by a million silken wings'. It was dusk when the insects started to rise up from the River Ebro in Spain for the last frantic hours of their lives. The date they emerge varies from year to year, depending on the temperature of the water and air conditions above, and on this river the peak can happen on any night between mid-August and mid-September (in this case, 13 September). The synchronization is astonishing. Having spent most of the year feeding under water, the nymphs (larvae) of *Ephoron virgo* ('virgin nymph') moult simultaneously into an adult flying stage. Meanwhile the male rapidly moults again into a second version with a longer tail for better flight stability and longer legs for grasping a female. The climax of the mating swarm may last just 15 minutes. Once mated, the females leave the frenzied mass to lay their eggs on the river, and then die. But here the mayflies have been lured to their death. Not only have the lights, like false moons, drawn them to the bridge, but the lamplight reflecting off the smooth, dark road surface resembles moonlight reflecting off the river. When light reflects off water it gets polarized, and *Ephoron virgo*, like all mayflies and many other aquatic insects, has evolved eyes to see polarized light to help it detect where to lay eggs. So having mated above the bridge, these females make the fatal mistake of laying their eggs on the road below. Since the sole purpose of the adult form of a mayfly is procreation, much of a generation can be lost on such a bridge of death.

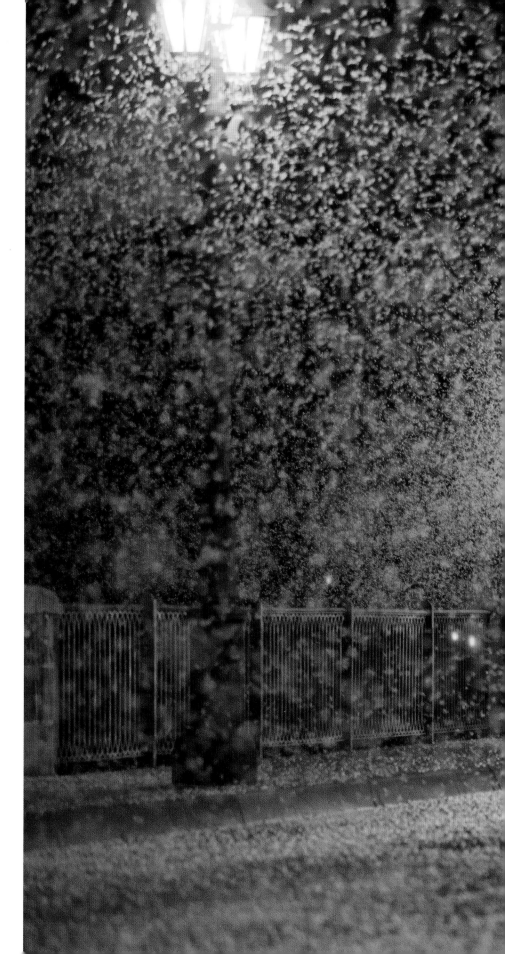

CREATIVE DINING
Brian Skerry

This is a remarkable aerial picture of a remarkable behaviour: a dining technique invented by bottlenose dolphins living on the lower Atlantic coast of Florida to catch fish in shallow water over mudflats, here in Florida Bay. Termed mud-ringing, one dolphin will swim at speed around a school of mullet, beating its tail against the seafloor to encircle the fish with a curtain of whipped-up mud. Terrified by the spreading curtain and approaching predator, the panicking fish leap out of the water, over the 'net' and into the open mouths of the dolphins waiting on the perimeter. It's an ingenious fishing strategy, developed by the dolphins to suit this particular environment and involves alliances and communications between the individuals involved. Not only do bottlenose dolphins identify themselves using distinct signature whistles, but they have been observed chattering among themselves while performing tricky tasks – which suggests that they might converse about tactics and timing. Mud-ring fishing is certainly a technique that has to be learned, with youngsters probably being educated by their mothers, and the likelihood is that the dolphin producing the ring, which doesn't get to eat the fish, is related to the ones which do.

TWIN HOPE
Diana Rebman

A portrait of mountain gorilla Kabatwa shows her suckling her six-month-old twins. That she looks tense is because the group's silverback leader, Munyinya, has just chased her off the patch of nettles she was feeding on. Munyinya was almost certainly the father of her boys, named Isango and Isangano at Rwanda's annual gorilla-naming festival. When they were born, in February 2011, they were only the fifth set of mountain gorilla twins ever recorded in Rwanda. Until they were strong enough to hold onto the fur on her back, Kabatwa had to carry them in her arms, which made walking and therefore keeping up with the group difficult. She continued to suckle them until they were about a year old, and they shared her leafy bed until they were four. Today, Isango and Isangano are healthy members of the group, adding to the population of this critically endangered species. Found only in Rwanda, Uganda and the Democratic Republic of the Congo, mountain gorillas now number about 880, of which 480 live in the Virunga Massif. But they remain at risk from habitat loss, human diseases, snares set for other animals, warring rebel factions active in their range and the implications of possible oil and gas exploration.

BREATH TAKING
Paul Nicklen

An all-male pod of narwhals surfaces to breathe among melting ice off Canada's Baffin Island in the High Arctic. Taking care not to hit each other with their long tusks, they take several breaths before diving again. Narwhals are tied to the Arctic ice, living underneath it for most of the year (they have no dorsal fin, presumably an adaptation to life under the ice). In the dark of winter they navigate using echolocation, always keeping in range of small cracks and leads that they can use when they surface to breathe. Migrating along set routes between their winter feeding areas under heavy pack ice and their coastal summering grounds, they feed on fish such as Greenland halibut and Arctic cod, squid and shrimps. They suck up their prey, having no chewing teeth, just two horizontal ones in the upper jaw. One of these erupts and grows through the lip of the male to form the spiralled tusk, which can be more than 3 metres (10 feet) long and is probably used for display or fighting (females occasionally grow much smaller tusks). In the Arctic summer, it's natural for some of the sea ice to melt, but as climate change causes the extent of it to shrink, so the narwhals' habitat shrinks, and in summer, they are now becoming more at risk from attack by killer whales.

BATTLE OF THE BULLS
Tim Fitzharris

Two bull northern elephant seals clash in a chest-to-chest fight, attempting to bite and slash each other with their canines. It's a boundary dispute between two veterans, as their scars show, and after just 10 minutes of biting and belching, the matter is sorted, and the trespasser backs off up the beach. Despite the blood, little damage is done, as they are protected by their thick, tough 'chest shields'. They hold neighbouring stretches of California beach territory at Point Piedras Blancas, a breeding colony for northern elephant seals, each fiercely guarding the harem of females lying within their estates. In early December, a few weeks before the pregnant females arrive to give birth on the beach, the males haul out to establish who gets beach territory – through threatening, chasing and fighting. Once dominance is established, a belching bout of throat trumpeting from a beachmaster can be enough to keep a younger male away. The beachmaster will then monopolize dozens of females that come ashore, mating with them after they give birth and preventing any other males getting access to them. But this requires constant vigilance. Bachelor males patrol in the surf zone waiting to grab a chance to nip up the beach and mate with a female on the edge of the harem, though 90 per cent of bachelors never actually father a pup. But the beach is not the only place to mate with females. Virgins may stay at sea for their first year, and other females take a year out of beach life, mating at sea, presumably giving at least some of the smaller males their 10 per cent chance to father a pup.

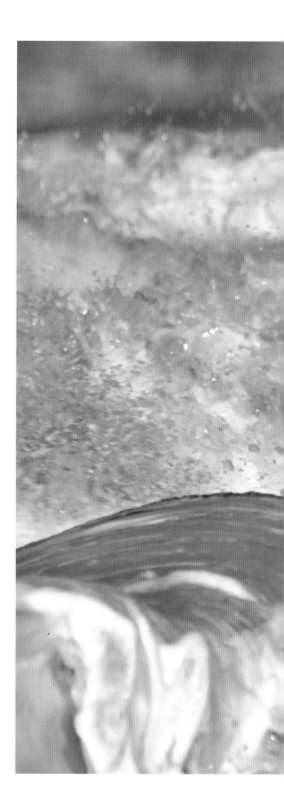

THE DANCE STAND
Todd Gustafson

Flamingos are famous for their ritualized and synchronized courtship dancing, but the locations where they dance are highly alkaline lakes. This is where they feed, on algae and brine shrimp, and if conditions are right, where they breed. The chief difficulty for a photographer wanting to portray the dance moves is to get close enough and then to single out a dance group from hundreds of movers and shoot them at eye level. Here the spot is at the edge of Kenya's Lake Nakuru, and the dancers are lesser flamingos. To show them at this angle, illuminated by the rising sun, required being there at dawn and lying on the mud. The tightly packed, fast-moving stand (the collective noun for flamingos) – possibly females as well as males – has posed balletically, feathers fluffed out, necks up and cherry-red beaks down. Dancing seems to be contagious, and group displays serve to synchronize reproduction among all the likely breeders in a huge flock so that, when the often unpredictable environmental conditions are right, nesting occurs en masse and the flamingo chicks hatch at the same time and are ready to move away together from the centre of the lake to feeding areas. This stand of flamingos has flushed extra pink, signifying that they are fit and well. In fact everything about them is pink, even their skin and insides, the colour coming from carotenoids in the algae they filter out of the surface water.

A LITTLE SWINGER
Thomas Vijayan

It's the end of the day, and a group of black-footed grey langurs are settling into a tree, ready for the night. But this infant has decided it's playtime. Swinging on the tails of two juveniles – seemingly unconcerned by his antics – he is making a characteristic play face. Eventually he falls off, but immediately climbs back up the tree and starts the game all over again, swinging another four times before finishing the session. His mother takes little notice of the activity, and the adolescents tolerate the attention-seeking acrobatics. Langurs are very social, and youngsters play as often as they can. But they do so less frequently when leaves and other plant material are not plentiful and feeding becomes all-important. Infants will play with older youngsters and subadults, but adults seldom join in. Acrobatic play is probably a way of practising tree-living and helps develop bone and muscle. But play also develops social bonds and communication skills, and the choice of play partner is important. Wrestling play is common and a way of sorting out who is dominant and learning the social dynamics of tolerance, cooperation and fairness; juveniles are very tolerant of infants and will let them take the lead and learn the signals and rules that stop play escalating into fighting. Young langurs even play at grooming, which is an essential part of social bonding. One theory about the need for play is that, among big-brained species, it is essential for brain development. Play is common only among mammals – though highly intelligent birds such as parrots and corvids (the crow family) also play – and there seems to be a relationship between the size of the brain and the amount of play. The suggestion is that the stimulation provided by play may actually mould the brain's architecture.

FLIGHT OF THE RAYS
Florian Schulz

You are looking at just a quarter of all the animals in a massive congregation of Munk's devil rays in shallow water in Mexico's Sea of Cortez, with as many again swimming under the surface. The photographer and his pilot, who had surveyed the area for 20 years, had never seen anything like it. All the rays were headed in the same direction, but what they were doing and why remains a mystery. The species itself was only described for science in 1987, and has been recorded in coastal waters of the eastern Pacific, from the Sea of Cortez south to Peru. Little is known about the ray's biology except that vast numbers, males and females, do gather, perhaps for mating, perhaps for feeding on planktonic crustaceans and small fish. They can swim fast, undulating their pectoral fins like wings, and regularly breach, often together in a sort of aerial show, exploding out of the water at speed, sometimes spinning or somersaulting, and then slapping back down on the surface. The most likely reason for jumping is communication, calling more rays into the aggregation. Unfortunately, like so many ray species, their habit of schooling makes them very vulnerable to fisheries or to being caught accidentally in drift-gillnets. Hundreds at a time are caught, and as they mature late and give birth to a single 'pup' after a long pregnancy, the numbers of these enigmatic and fascinating migratory rays has to be declining.

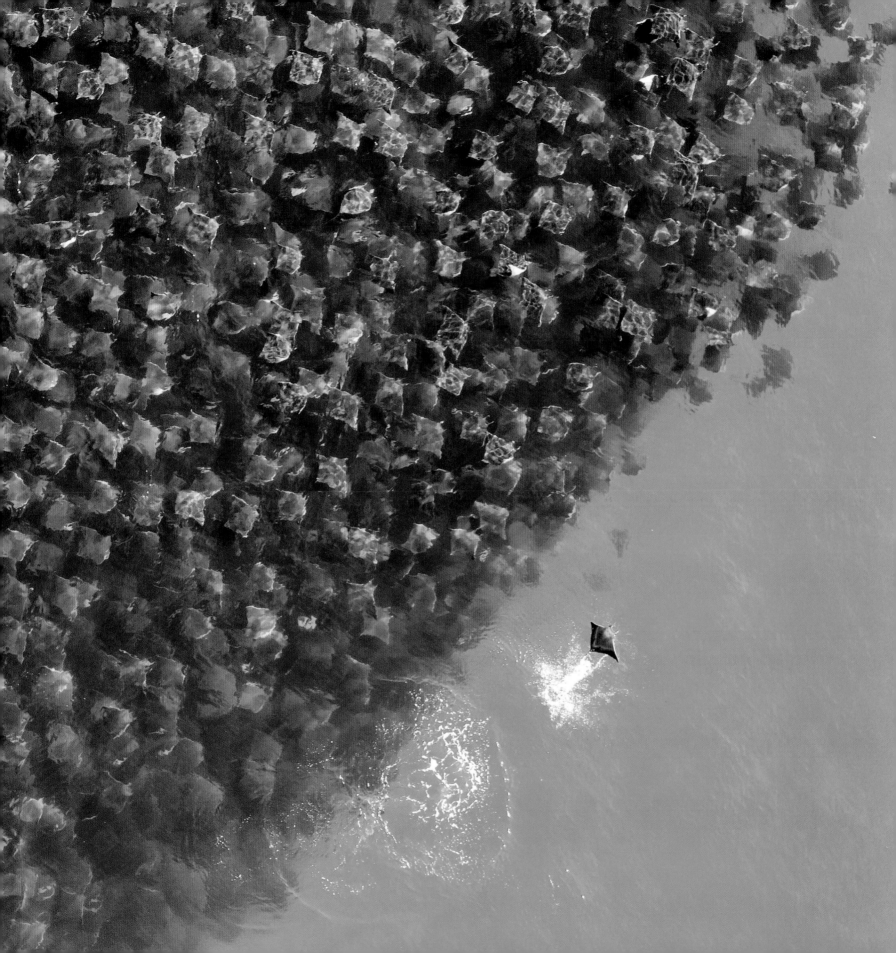

MARCH OF THE CRABS
Pascal Kobeh

Every year, as winter draws in, something triggers the Australian majid spider crabs to start marching. It could be the change in water temperature or maybe a moon cycle, no one knows. And indeed, no one knows exactly where they live and what they do the rest of the year, except that, at some point from April to June, they head for shallow water. Thousands of them, sometimes hundreds of thousands, start to meet up as they march towards sheltered bays off Victoria, South Australia and Tasmania. As they reach the shallows, some of them may form piles more than a metre high, perhaps an exaggerated safety-in-numbers exercise if they feel threatened. They come to the bays for a mass moult. Their shells are effectively their skeletons – they have no bones – and to grow and to replace any breakages, they need to replace their exoskeleton with a new one. It's an exercise that needs calm water, but it makes them vulnerable, hence the safety-in-numbers strategy. Not only can it take up to half an hour for a spider crab to extract its legs and body, but it emerges with a soft new shell that is conspicuously orange, lacking the camouflage that its old shell had acquired, often including an encrustation of marine organisms. Most of the spider crabs moult within a week, and once their new shells have hardened, they depart, leaving the seafloor littered with their old ones.

BUBBLE TALK
Paul Nicklen

It's open to interpretation what the leopard seal is communicating to Paul Nicklen with its bubble-blowing and gift of a penguin. Little is known about its behaviour, and few people have had such an intimate encounter. Setting out to observe the reaction of leopard seals to his presence in the water, aiming to dispel their fearsome reputation, Paul chose to dive off the coast of the Antarctic Peninsula island of Anvers at a time when young penguins were leaving their colonies – easy pickings for leopard seals. When he first entered the water, this huge female – 3.5 metres (12 feet) long – swam straight over and lunged at him, mouth wide open, almost touching his mask. He took this as a dominance threat. Indeed, given that her head was larger than that of a grizzly bear, it was nerve-racking. But the next time he entered the water, she swam over holding a live penguin by its feet. Releasing it, she let it swim towards him, caught it and then released it again, repeating this several times before eating it in front of him. Several more penguins were offered that day. On the following day, she came over with another penguin. But this time she killed it and presented him with the body. When Paul showed no indication of doing anything with it, she started blowing bubbles at him, which he took as a sign of frustration. Without doubt, she wanted to communicate and had no intention of harming him. There are records of occasional unprovoked attacks on people, including a British scientist who was drowned, but Paul believes his experiences with leopard seals show that, like all mammal predators, they are individuals and not necessarily aggressive if treated with respect.

COLD-BLOODED KILLING
Alejandro Prieto

It was the noise that alerted the photographer: a hammering and a mighty splashing. He was on a remote beach on the Pacific coast of Costa Rica, looking for bull sharks, which had been sighted hunting close to shore. But what he saw was a huge American crocodile holding a large green turtle by its shell. The turtle was thrashing around, trying to escape. Once out of the water, and with a fast flick of its jaws, the crocodile grabbed the turtle by a flipper and then by its head – the surest way of killing it, rather than breaking through the shell. It then dragged its prey back into the surf, heading towards the river estuary, presumably to haul out there and crunch through the turtle. Crocodilians have the strongest bite of any animal on Earth, capable of smashing through turtle shells once they can get their jaws around them. The secret of a crocodilian's bite is its second jaw joint, which helps to distribute the force when it crunches down and stops its jaw twisting as it feeds. Indeed, sea turtles may well have been a favourite dish of the huge prehistoric crocodiles that hunted offshore, and today's alligators and crocodiles regularly eat freshwater turtles. Saltwater crocs have also been recorded actively hunting nesting turtles, following their tracks up the beach to catch them laying. But for a sea turtle, the greatest threats it faces are not crocodiles or even the sharks that prey on them but those of human origin: nets, boats, pollution and waste, especially plastics.

MIRACLE BIRTH
Adrian Hepworth

Out of the sea, a yellow-bellied sea snake is helpless. It has a body adapted for swimming – compressed and with a paddle-like tail – so it can't crawl like a land snake. And if stranded in the sun, it will die. In this case, it was early morning when the photographer saw the mother snake, washed up after a storm, apparently dead, on a beach on the Pacific coast of Costa Rica. Just five minutes later, walking back the same way, he was astonished to see that she was not only alive but had given birth. As the tide rushed in, he watched as the baby wriggled energetically next to its fatigued mother. Once a wave reached it, the baby shot away, but the mother took a while to push through the incoming tide. Picking her up would have been risky. There is no danger of being bitten by a yellow-bellied sea snake in the sea – it has a small mouth and small fangs – but if handled the wrong way, it can bite, and its venom, used to immobilize fish, can be deadly to humans. And there was a danger that handling her with a stick might have damaged her flaccid body. She may also still have had young inside her (there are normally two to six in a litter). The yellow-bellied sea snake is the most truly ocean-living snake and the widest ranging, found in the warm waters of the Indo-Pacific and eastern Pacific oceans. Not a lot is known about its behaviour, but the use of black and yellow warning colours indicates that it is toxic. Its ancestors were land snakes, and it still needs to drink fresh water, which it does off the surface of the sea after rain, and this may explain why they are sometimes found offshore, near estuaries or, after a drought, where there is the likelihood of a rainstorm.

CRÈCHE CALL-UP
Darío Podestá

A young Patagonian mara presents its bottom to its mother so she can sniff its identity. She has whistled the brood out of the communal den on the southern plains of Argentina and is checking which are her twins. Once she's identified them, by size as well as scent, she will suckle them nearby, fending off the other hungry youngsters. She's accompanied by her mate, who sticks close to her and keeps watch for predators. The rest of the crèche have to wait for the daily visit of their own parents. They keep close to the den and go to ground immediately if an adult warns that an eagle or fox is about. More than 20 pairs have been known to den together. The more that do so, the greater the survival rate of the young, perhaps because there will almost always be one pair in attendance and therefore adults on guard. And if any young are orphaned, they have a chance of suckling from the other mothers. Such crèche denning is unique among mammals. On the plains, it's a way to keep the youngsters safe for their first few weeks, until they are old enough to follow their parents on feeding forays. Patagonian maras are giant rodents – that is, related to guinea pigs, not rabbits – found only in Argentina. Not only is their breeding system unusual, but unlike most other mammals, they pair for life, the male following the female everywhere as they search for grazing, scanning for predators as she feeds. They are fast runners and need to be as they are prey to predators from foxes to pumas. Like gazelles, if they spot a predator, they use an exaggerated bounding gait that advertises that they are fit and not worth the chase.

THE SNAKE EAGLE FAMILY
José B Ruiz

The nest of these short-toed snake eagles is high up in an Aleppo pine tree in Alicante, Spain, overlooking the dry, shrubby territory where the birds hunt. The female is titivating the nest, while the male is about to regurgitate a long ladder snake for their chick. Even though the chick is still small, it will swallow the snake in one go. Snake eagles only lay one egg (most birds of prey lay two, the second egg as an insurance policy), devoting much attention to their single chick and continuing to feed it for several weeks after it has fledged. In this region, they eat chiefly Montpellier snakes followed by ladder snakes. Montpellier snakes eat other snakes, and so by preying on them, the eagles inadvertently help maintain snake diversity. But being specialist snake-eaters, their survival is also tied to them. Though snakes are legally protected in Spain, they are still persecuted and are also killed while lying on the roads warming up, and their numbers are declining with loss of their habitat. It's not known how much this is affecting the snake eagles, especially as eagles are long-lived, but they may well be declining. Once a chick is fully independent, the adults will migrate to spend the winter in Africa south of the Sahara (the youngster will follow them a bit later). Occasionally, some eagles stay in Spain, but presumably only in areas where snakes remain active in winter.

A MAELSTROM OF MILLIONS
Ewald Neffe

It's sunset in February, and already every branch is thick with bramblings while yet more and more pour into the forest from all directions, swirling overhead like starlings before finding a spot to roost. The air vibrates with the whirr of wings and the churring rattle of the birds' communication. There is also a constant rain of droppings. A handful of predators, sparrowhawks and peregrines, try to catch the odd bird, but there is definitely safety in numbers for the bramblings by flocking in such a multitude. By the time it's dark, four million will have swept in to gather in a tiny section of this southern Austrian forest. They are migrants in search of food in advance of the onset of winter. In Europe, these chaffinch-sized, mainly seed-eating finches move south and west in search of tree seeds. Beechmast is a favourite, and if there is a good crop in a particular region, they might roost in the same spot until the mast runs out or snow covers it over. Winter flocks tend to be predominantly one sex, with young birds moving farther south than adults, perhaps because of competition for food. In this instance, almost all the birds were males. Though flocks more often number in their thousands, massed migrations a million strong are not uncommon, and one recorded in southwest France in 1967 numbered 20 million. In Britain and Ireland, however, the total wintering population from Scandinavia and Russia is seldom more than a million, and so roosts are comparatively small.

EXTREME FORAGING

Ron McCombe

A female red grouse stands on tiptoe to pluck the last evergreen leaves and seeds from an exposed stand of heather. She has been forced down from the high moorland in southern Scotland by an unusually long (three-week) period of snowfall, which has buried most of the heather. The red grouse is the British race of willow grouse, but unlike its Arctic tundra relatives, it feeds mainly on *Calluna* heather, especially in winter. Nor does it moult into a white winter plumage, presumably because winters are not usually long enough to merit such a major investment in camouflage. Cold is seldom a problem. A double layer of dense plumage provides enough warmth, while its legs down to its toes are covered in feathers. And to protect itself from a bitter wind, a red grouse digs a burrow into the snow, which acts as insulation. But when snow is heavy and covers the heather, starvation is the main threat. With its long gut and special enzymes and gut microflora, the red grouse can extract the maximum amount of protein and nutrients, such as nitrogen and phosphorus, from any heather it finds, though such a high-fibre diet requires it to eat grit to help break down tough evergreen leaves, and it also has to detoxify the tannins in them. In spring, though, the grouse feast on the nutritious shoot tips of heather, which the females in particular rely on to produce a healthy clutch of eggs.

THE ASSASSIN
Steve Mills

Snow not only provided the perfect backdrop for a perfect picture but also the ideal opportunity for a merlin to catch a snipe. In early December 2010, two weeks of severe winter weather had frozen the ground throughout much of the UK. Here in North Yorkshire, there were few areas of soft ground where common snipe could probe for worms, insects and other invertebrates. Normally, when feeding in the open, a snipe relies on its plumage camouflage. If it senses an attack, it will burst from cover in a rapid zigzag flight that makes it difficult for a hawk to catch it. But here the bird was forced to feed on a precious area of snow-free grass at the edge of a field where its camouflage was useless, and it just didn't sense the surprise attack. The young female merlin may have staked out the patch, as indeed, the photographer had, and was on a nearby perch. She swooped in fast and low and pinioned the snipe to the ground, then killed it with half a dozen blows to the back of the head. A merlin is slightly smaller than a kestrel, and its diet is more normally birds the size of pipits and skylarks, together with a small proportion of insects, such as moths and dragonflies, as well as the occasional rodent. Females, though, are larger than males and will target prey as big as snipe, and in winter when many smaller birds such as swallows have migrated south, snipe are a sought-after catch, their numbers swelled by migrants escaping the harsh winters of northern Europe.

INSIDE JOB

Charlie Hamilton James

Rüppell's griffon vultures dominate the feast on the plains of Tanzania's Serengeti National Park. Ranked behind are the smaller (dark-billed) white-backed vultures, waiting their turn to feed on what's left of the zebra carcass. In the background (far right) is a large marabou stork digesting its fill. African vultures hunt for dead animals by sight rather than smell, soaring on thermals and keeping an eye on the movements of each other and other scavengers, including eagles, which may well find a body first and eat the soft, accessible eyes and tongue. If an animal has died of natural causes – a wildebeest on migration, for example – the smaller vultures have to rely on the arrival of carnivores or a huge lappet-faced vulture with its big, hooked bill to rip open a corpse. This zebra was killed by a lion and so was open to all, but the risk for the vultures was that it might have been laced with the pesticide Carbofuran, now frequently used by pastoralists to poison kills or as bait to kill predators. A vulture's stomach acid will destroy dangerous bacteria but not pesticides, and in the past decade, thousands of vultures have died from poisoning. Ivory poachers are also poisoning vultures to stop them acting as beacons to reveal elephant kills. Other threats to these slow-breeding birds – nature's essential clean-up team – include hunting them for the trade in traditional medicines and collisions with wind turbines. The crash of African populations of vultures is such that numbers of seven key species have fallen by more than 60 per cent over the past 30 years and 6 species could by now be critically endangered.

GARBAGE PICKING
Jasper Doest

Spaced out over the colourful mountain of plastic refuse bags and other garbage on a landfill site in Andalucía, southern Spain, are adult and adolescent white storks, looking for food. Unfortunately, the deceptive colours and shapes of garbage cause them to ingest potentially lethal scraps, mainly plastics and also rubber bands, which mimic earthworms. And in the breeding season, some of this material will be fed to chicks. In Andalucía, the attraction of such landfill sites is that they are available all year round and are constantly topped up, and some populations are now reliant on rotting food remains rather than consuming a varied natural diet – amphibians, rodents, young birds, fish, crayfish, insects and earthworms. Indeed, the breeding and wintering distribution of these storks as well as their migration flyways have been moulded by the distribution of open dumps. What has worried biologists is the amount of plastics swallowed by young birds and the level of contaminants that build up in the storks' eggs, which could affect the health of future generations. On the other hand, the numbers of Andalucía's white storks have increased just because of the year-round refuse. So now that dumps are being closed – the rubbish being recycled, buried as landfill or incinerated – as required by legislation, there is greater concern that the Spanish population of these wonderfully opportunist birds will once again drop to worrying levels, given the loss of their natural feeding areas to drainage and agricultural intensification.

DADDY LONG-LEGS ON GUARD
Jordi Chias

This spiny grouping of animals, in the sea off the Canary Islands, is composed of a pair of long-spined sea urchins and a pair of eastern Atlantic arrow crabs. These crabs are popular subjects with underwater photographers, because of their eight spider-like walking legs, but this picture tells a whole story about their behaviour and lifestyle. The crabs are using the sea urchins as a back-up defence against predators, though presumably their prickly arrow-shaped heads help deter some fish from trying to eat them. The male is standing astride the female to guard her and her orange eggs not only against predators but also against the attention of other males. She has just moulted, and when her new shell was still soft, he inserted his sperm packet into her genital opening. Though it is likely she will use his sperm to fertilize her eggs, there is still the danger of a rival trying to insert a second sperm plug, which is why the male remains on guard. After she has used the sperm for fertilization, she will continue to hold the clutch (visible under her abdomen) with her abdominal appendages until her young hatch and swim away.

FLYING LEAP
Paul Nicklen

An emperor penguin, its belly full of food, shoots out of the water and aims for the ice. The speed of propulsion has been aided by an extra streamlining of its torpedo-shaped body through a lubricating film of bubbles that it released from its feathers, enabling it to double its speed as it rocketed to the surface. Such speed is needed both to clear a good few feet of ice and to escape being ambushed by any leopard seal lying in wait at the ice hole. Relying also on safety in numbers, it has returned in a wave of hundreds, which are exiting in fast succession. Its belly is full of extra cushioning for a thumping landing – krill, fish or squid – which it will regurgitate to its chick waiting at the colony some 10 kilometres (6 miles) further in on the frozen western Ross Sea on the Antarctic coast. The size of a small, overweight child, an emperor penguin makes an ideal meal for a leopard seal, which will lurk around ice holes and landing areas that the emperors use and where their newly fledged young will take their first dives in late summer. But once these biggest of penguins are back in open water, they can outmanoeuvre a leopard seal and are relatively safe – unless they meet a pod of killer whales.

INDEX

PHOTOGRAPHERS

Adrian Bailey, South Africa
20, 66
www.baileyphotos.com

Jim Brandenburg, USA
10
www.jimbrandenburg.com

Jordi Chias, Spain
122
www.uwaterphoto.com

Jasper Doest, The Netherlands
120
www.doest-photography.com

Göran Ehlmé, Sweden
54
goran@waterproof.eu

Kai Fagerström, Finland
64
www.kaifagerstrom.fi

Tim Fitzharris, USA
92
www.timfitzharris.com

Angelo Gandolfi, Italy
52
gandolfiangelo@gmail.com

Howie Garber, USA
34
www.wanderlustimages.com

Linc Gasking, USA
22

Mike Gillam, Australia
82
www.vanishingpointgallery.com.au

Axel Gomille, Germany
14
www.axelgomille.com

Sergey Gorshkov, Russia
78
www.gorshkov-photo.com

Marcel Gubern, Spain
60
www.oceanzoom.com

Todd Gustafson, USA
94
www.gustafsonphotosafari.com

Don Gutoski, Canada
28
www.pbase.com/wilddon

Charlie Hamilton James, UK
50, 118
www.charliehamiltonjames.com

Bill Harbin, USA
16
wpharb@comcast.net

Adrian Hepworth, UK
8, 106
www.adrianhepworth.com

David Herasimtschuk, USA
18, 58
www.davidherasimtschuk.com

Pascal Kobeh, France
100
www.scuba-photos.com

Zig Koch, Brazil
74
www.zigkoch.com.br

Tim Laman, USA
56
www.timlaman.com

Ron McCombe, UK
114
www.wildlife-photography.uk.com

José Antonio Martínez, Spain
84
www.joseantoniomartinez.com

Bence Máté, Hungary
1, 4, 6, 62
www.matebence.hu

Thomas D Mangelsen, USA
2
www.mangelsen.com

Steve Mills, UK
116
www.stevemills-birdphotography.com

Mark W Moffett, USA
42
Minden Pictures/FLPA
www.flpa-images.co.uk

Ewald Neffe, Austria
112
www.ewaldneffe.com

Paul Nicklen, Canada
90, 102, 124
www.paulnicklen.com
National Geographic Creative
www.natgeocreative.com

Mark Payne-Gill, UK
70
www.mpgfilms.co.uk

Eric Pierre, France
26
www.boreal-lights.com

Darío Podestá, Argentina
108
www.dariopodesta.com

Benjam Pöntinen, Finland
48
www.pontinen.fi

Louis-Marie Préau, France
46
www.louismariepreau.com

Alejandro Prieto, Mexico
104
www.alejandroprietophotography.com

Udayan Rao Pawar, India
24
udayanraopawar17@gmail.com

A FIREFLY BOOK

Diana Rebman, USA
88
www.dianarebmanphotography.com

Jenny E Ross, USA
32
www.jennyross.com

Getty Images
www.gettyimages.co.uk

Andy Rouse, UK
38, 68
www.andyrouse.co.uk

José B Ruiz, Spain
110
www.josebruiz.com

Nature Picture Library
www.naturepl.com

Joel Sartore, USA
30
www.joelsartore.com

Kevin Schafer, USA
80
www.kevinschafer.com

Florian Schulz, Germany
98
www.visionsofthewild.com

Anup Shah, UK
72
www.shahrogersphotography.com

Igor Shpilenok, Russia
76
www.shpilenok.com

Brian Skerry, USA
86
www.brianskerry.com

Paul Souders, USA
44
www.worldfoto.com

Maria Stenzel, USA
12
www.mariastenzel.photoshelter.com

National Geographic Creative
www.natgeocreative.com

Klaus Tamm, Germany
40
www.tamm-photography.com

Thomas Vijayan, India
96
www.thomasvijayan.com

Christian Ziegler, Germany/USA
36
www.naturphoto.de

Published by Firefly Books Ltd. 2017

Copyright © 2016 The Trustees of the Natural History Museum, London
Photography © 2017 The individual photographers

First printing

Publisher Cataloging-in-Publication Data (U.S.)

A CIP record for this title is available from the Library of Congress

Library and Archives Canada Cataloguing in Publication

A CIP record for this title is available from Library and Archives Canada

Published in the United States by
Firefly Books (U.S.) Inc.
P.O. Box 1338, Ellicott Station
Buffalo, New York 14205

Published in Canada by
Firefly Books Ltd.
50 Staples Avenue, Unit 1
Richmond Hill, Ontario L4B 0A7

Printed in China

First published by the
Natural History Museum,
Cromwell Road, London
SW7 5BD

Editor and author
Rosamund Kidman Cox
Designer
Bobby Birchall,
Bobby&Co Design
Image grading
Stephen Johnson
www.copyrightimage.com